Bangor Garth Pier opened in 1896. It was built using cast-iron columns and screw piles, with wrought iron and steel for other structural members; it reaches two-thirds of the way across the Menai Strait. The ornamental ironwork and series of kiosks dotted along its 1550 foot (472 metre) length give it the appearance of an early Victorian pier.

Piers and Other Seaside Architecture

Lynn F. Pearson

A Shire book

Published in 2002 by Shire Publications Ltd,
Cromwell House, Church Street, Princes Risborough,
Buckinghamshire HP27 9AA, UK.
(Website: www.shirebooks.co.uk)

British Library Cataloguing in Publication Data:
Pearson, Lynn F.
Piers and other seaside architecture. – (A Shire album; 406)
1. Piers – Great Britain
2. Seaside architecture – Great Britain
3. Pavilions – Great Britain
I. Title
725.8'0941
ISBN 0 7478 0539 3

Front cover: *Llandudno Pier in north Wales was built in 1876–7 and designed by Sir James Brunlees; it is a fine example of a relatively unchanged Victorian pier, with its series of original kiosks and abundant ironwork. The octagonal-ended pierhead pavilion was added in 1905, while the earlier shore-end winter garden pavilion (1883–4), next to the Grand Hotel, was demolished following a fire in 1994; the surviving remains include an iron staircase left hanging in mid air.*

Back cover: *Brighton Palace Pier, East Sussex.*

ACKNOWLEDGEMENTS
The author is very grateful to the following for their assistance, in many and various ways, with this book: Penny Beckett, Martin Easdown, Gwyn and Marie Evans, John Garlick, Tony and Kathy Herbert, Sue Hudson, Biddy Macfarlane, Geoff Maggs, Chris and Mick Paxton, Jim and Margaret Perry, John Rotheroe, Derek Sugden, Jane Webb, Mark Whyman, Blackpool Borough Council, Blackpool Pleasure Beach, the Friends of the Midland Hotel, the Museum of London, the National Piers Society, Southern Water, Spence Associates and the Tiles and Architectural Ceramics Society. Illustrations are acknowledged as follows: courtesy of Blackpool Pleasure Beach, page 32 (bottom); courtesy of Tony Herbert, page 37 (bottom left); courtesy of Cadbury Lamb, page 6 (bottom); courtesy of the Marlinova Collection, page 29 (both); courtesy of the Museum of London, page 20 (top); courtesy of Snøhetta + Spence, page 40 (bottom); courtesy of Southern Water, page 44 (centre and bottom); author's collection, pages 9 (top), 11 (bottom), 13 (top), 14 (bottom), 15 (top), 17 (bottom), 27, 28 (top), 31 (top). All other photographs were taken by the author.

Printed in Malta by Gutenberg Press Limited, Gudja Road,
Tarxien PLA 19, Malta.

Contents

Saltburn Pier was designed by John Anderson and opened in 1869; a bandstand was added to the pierhead in 1885 but lost after a ship collided with the pier in 1924. The entrance buildings date from the latter part of the 1920s. The pier was breached for defence purposes during the Second World War, then rebuilt in 1947.

Pleasure piers and pavilions

Above: *Bangor
Garth Pier was
designed by the
engineer John James
Webster and built by
the contractor Alfred
Thorne; the same
combination
successfully
completed the
construction of a pier
at Dover in 1893.
The design of
Bangor's
promenading pier is
enlivened by
oriental-style toll-
houses and kiosks,
the latter supported
on braced towers at
intervals of 250 feet
(76 metres).*

The concept of the pleasure pier originated out of the sheer difficulty of reaching the seaside watering places that had become popular during the late eighteenth and early nineteenth centuries.

Prospective visitors initially had the choice of expensive stagecoaches or slow sailing-boats, but the introduction of quick, reliable steam passenger vessels around 1815 transformed the journey to nascent resorts such as Weymouth and Margate. There remained, however, the problem of transferring passengers from ship to shore where docking facilities were inadequate. Travelling to the Isle of Wight by ship to Ryde involved a long walk across a wet, sandy beach when arriving at low tide. This was seen by the townspeople as a deterrent to visitors, and a landing pier was built in 1813–14. Although little more than a long, narrow wooden jetty – it was 1740 feet (530 metres) in length but only 12 feet (3.7 metres) wide – and completely bereft of any decorative features, the pier was so successful that it was repeatedly extended and the pierhead enlarged during the following two decades.

As with the piers built at Brighton, Leith, Southend, Walton on the Naze, Herne Bay, Southampton and Deal during the 1820s and 1830s, landing charges for goods and passengers provided the income for Ryde Pier, but during the 1840s it became obvious that piers were also being used for reasons other than the purely functional. Visitors, attracted to the sea but repelled by its danger, could promenade along the pier in complete safety and still enjoy a *frisson* of excitement caused by the proximity of the waves below.

The noise of the sea, the therapeutic sea breeze, the activities on the landing-stage, the view of ships and shore, all made the pier a unique resort attraction, and one that could clearly be made more profitable, particularly as the expansion of the railway network during the 1840s and 1850s brought the seaside within easy reach of many more people than before. The addition of a toll-house, turnstiles, or both, at the shore end of the pier allowed a charge to be made for promenading, either daily or by means of a season ticket. A classical toll-house was installed at the entrance to Ryde Pier before 1840, and later piers utilised picturesque or oriental styles for these small but significant structures.

Facilities for promenaders slowly improved: a further pavilion was added to Ryde Pier in 1842, while the 200 foot (61 metre) length of the Royal Terrace Pier (1843–5) at Milton-on-Thames, near Gravesend, was both illuminated and covered. Great Yarmouth's Wellington Pier (1853) included a promenading platform at its

Southport Pier, opened in 1860, was one of the earliest pleasure piers. Its maximum length after extensions during the 1860s was 4380 feet (1335 metres), but damage by storm and fire has reduced this to 3633 feet (1107 metres), still making Southport the second-longest pier in Britain after Southend. The pier was threatened with demolition in the early 1990s but its first phase of restoration was due for completion in 2002.

Colwyn Bay Victoria Pier was designed by the specialist seaside architects Mangnall & Littlewoods of Manchester and built in 1899–1900. The prefabricated iron substructure carried a 2500-seat pavilion, which burned down in 1922. Its replacement was also destroyed by fire in 1933, although another pavilion opened the following year. The pier has been under threat of closure or demolition since the 1970s; although it reopened in 1995, further restoration is still required.

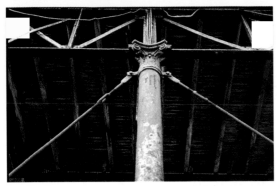

Every iron column at Colwyn Bay Victoria Pier is stamped with the ironfounders' name – the Widnes Foundry Company – and an individual column number; the decorative capitals support a horizontal lattice girder. The Widnes Foundry Company also supplied ironwork for Morecambe West End Pier (1893–6), Mumbles Pier (1897–8) and Great Yarmouth Britannia Pier (1900–1).

seaward end, and Southport Pier (1859–60) was built primarily as a promenading pier, although much of its success resulted from its position on a regular paddle-steamer route.

Southport was also unusual because of its building method, which avoided pile-driving. In contrast to harbour piers, which were normally built from solid masonry, seaside piers were generally constructed of piles connected by a lattice framework; this was cheaper than masonry and less resistant to water movement. A seaside pier usually comprised upright piles, girders connecting the piles at their top ends, decking above the horizontal girders, and struts and bracing members linking the previous three. Initially, piers were built wholly of timber, the piles being driven into the seabed by piling engines, but engineers found that the timbers were attacked by marine worms and borers and so from the mid 1840s began to use cast-iron piles. At first the piles were square in cross-section but these were soon replaced with circular castings, which offered less water resistance, and the invention of screw piles provided a further improvement. Screw piles, fitted with blades enabling them to be screwed directly into the ground, were first used at Margate Pier (1853–7) by the civil engineer Eugenius Birch (1818–84), the foremost of the Victorian pier designers. Neither screw nor driven piles were used for Southport and its two 1860s extensions, but a method known as jetting, in which water was pumped down through a tube within the pile, agitating the sand beneath and allowing the pile to sink into place.

Clevedon Pier's eight arches, each spanning 100 feet (30 metres), were created using Barlow rails, a patent wrought-iron rail section, riveted to curved iron flats. The pier opened in 1869 and the pavilion, by Melton Iron Works, was added in 1894. To accommodate the wide tidal range (up to 49 feet or 15 metres), flights of steps ran up through the pierhead so that passengers could alight at any state of the tide.

Left: *Birnbeck Pier (1864–7) at Weston-super-Mare links the mainland with rocky Birnbeck Island, then continues out into the Bristol Channel with a 250 foot (76 metre) wooden landing-stage added in 1872. The pier was lit by gas lamps and was immediately popular, with 120,000 people paying the one penny admission charge in the first three months. A pavilion was added to the pier in 1884.*

Right: *Eugenius Birch worked with the Glasgow contractor Robert Laidlaw on the construction of Blackpool North Pier (1862–3), and the pair went on to build piers at Brighton (West), Deal, Lytham and Hastings. The North Pier's tiny kiosks were part of an ornate design that included intricate wrought ironwork forming continuous seating along the pier sides.*

Initially, timber was used for the horizontal members, but this was soon replaced by wrought iron, its torsional strength making it better suited for this purpose than cast iron; cheap steel replaced wrought iron in the late nineteenth century. Girders tended to be either solid I-section or latticework in varying formats; the latter were lighter and were inherently decorative. Along with lattice girders, different arrangements of columns and struts (taking compression) and ties (in tension) could result in a highly ornate and elegant substructure. The delicate, curving ironwork of Clevedon Pier (1867–9) is the most memorable in this respect, but the complex system of angled bracing used by Birch at Weston-super-Mare's Birnbeck Pier (1864–7) is also very attractive. Decking was normally made of wood, the timbers often being laid herring-bone style and cambered.

The piers of the 1860s were transitional, being neither mere landing-stages nor pure pleasure piers; their architects tentatively embarked on the addition of decorative features designed to pull in

7

The Royal Pavilion, Brighton, is an extravagant Indian-style palace created in 1815–23 from the Marine Pavilion (1787) by the architect John Nash for George, Prince of Wales, who was first Prince Regent (1811–20), then King George IV (1820–30). The artist-designers Frederick Crace and Robert Jones produced the richly oriental interior decoration. The Pavilion was sold to Brighton by Queen Victoria in 1850.

the potential crowds. Two piers designed by Eugenius Birch included a series of substantial kiosks along the length of the pier: at Blackpool North (1862–3) these were Italianate in style, while in Brighton Birch took his lead from the Royal Pavilion (which had been transformed into an exotic domed palace for the Prince Regent in 1815–23) and selected an oriental theme for the West Pier (1863–6).

Shops, first seen as an addition to Brighton's Chain Pier (1822–3) in the 1830s, reappeared on Rhyl Pier (1867) along with a restaurant and a bandstand, while the Birch-designed New Brighton Pier (1866–7) in the Wirral featured restaurants and a central saloon with an observation tower. A few of the longer piers also incorporated passenger railways, some of which began as the contractors' own lines, with propulsion originally provided by hand or horse. Southport's cable-operated passenger tramway opened in 1865.

It was still the case that sixteen of the twenty-two piers begun during the 1860s were plain landing-stages with the minimum of buildings. Placing more emphasis on pier entertainment required the companies which promoted pier building to put up increased capital, although with the incentive of enhanced financial returns from promenaders spending extra time and money on the pier.

Promoting a pier was not without its drawbacks. Consent for the construction of a pier had to be obtained from Parliament by means of a parliamentary bill, and from the foreshore authority, which was often the Board of Trade but which could involve local lords of the manor and other bodies. Once consent had been obtained, building took place in an environment surpassed in difficulty only by the construction of rock lighthouses and, even when the pier was complete, the unique combination of dangers from waves, wind, fire and collision led to the loss of many pier pavilions and sections of pier. Of course, not all proposals for piers resulted in a successful conclusion: Lowestoft New Pier (1899), intended to extend from Britain's most easterly point, was never built, while the delightfully domed Whitley Bay Pier (1908) reached the design stage but no further and the resort remained pierless.

Finance for pier promotion might be obtained from local residents and businesses, or regionally, often from urban centres in the catchment area of the resort, or sometimes nationally. The most

Eugenius Birch included this domed and arcaded seaward-end 2000-seat pavilion in his original design for the 910 foot (277 metre) Hastings Pier. The pavilion was burnt out in 1917 and replaced by a less elaborate structure in 1922. The pier's substructure comprises cast-iron columns on screw piles with lattice girders above.

common means of raising capital was via the joint stock company, which could be formed by a minimum of seven people, but it was the advent of limited liability (following law changes after 1855) that led to a boom in small share ownership, and many more resort-based pier and entertainment companies were formed from the 1860s onward.

The pier that heralded the exuberant superstructure of pavilions and kiosks now associated with the Victorian and Edwardian seaside was Hastings Pier (1869–72), which was the first to have a

Aberystwyth Royal Pier (1864–5) was built by Eugenius Birch and the contractor J. E. Dowson, with whom Birch also worked on New Brighton, Eastbourne and Scarborough piers. Dowson died in the late 1860s before the last two were completed and the Stockton firm Head Wrightson took over the contracts. The Aberystwyth pavilion dates from 1896; in the distance is the cliff railway, which opened in the same year.

Llandudno Pier is notable for the fact that it is a dog-leg in plan rather than straight, and for its lavish use of ornamental wrought and cast ironwork in balustrades and railings. Each of the octagonal kiosks has ironwork of unusual design beneath its overhanging roof.

Above left: *Before its development as a resort from the 1850s, Llandudno had been little more than a hamlet lying between the Great Orme and Little Orme headlands. The Mostyn Estate built wide promenades and impressive terraces, while the pier was erected by the Llandudno Pier Company in 1876–7. The contractor was John Dixon, a railway builder who had been responsible for shipping Cleopatra's Needle to England and had worked with James Brunlees on Southport Pier, perfecting the jetting system of construction.*

substantial pavilion (with accommodation for 2000 people) included as an integral part of the original plan. Both pier and oriental-style pavilion were designed by Eugenius Birch, who built a total of fourteen piers in England and Wales, seven of which survive: Aberystwyth, Blackpool North, Bournemouth, Brighton West, Eastbourne, Hastings and Weston-super-Mare (Birnbeck). His pioneering use of the oriental style during the 1860s instigated its revival, and it eventually became the archetypal English seaside architectural style. Birch's huge contribution to the seaside included the first recreational aquarium (Brighton, 1872) and Scarborough Aquarium (1877), the latter being an extravagantly decorative Indo-Moorish structure, which was demolished in 1968 to make way for a car park.

Another notable pier designer was Sir James Brunlees (1816–92), a civil engineer who generally worked on railways and docks. He used the unusual water-jetting system to construct Southport Pier and was also responsible for the piers at Llandudno (1876–7) and Rhyl (1867); he built the second Southend Pier in 1888–90. Joseph William Wilson (1829–98), an engineer who had worked on the Crystal Palace, built four piers during 1864–73: Bognor Regis, Teignmouth, Westward Ho! and Hunstanton. John James Webster (1845–1914), an engineer who specialised in bridge construction, designed piers at Dover, Bangor and Minehead in 1892–1901.

The ironwork of the railings at Colwyn Bay's Victoria Pier, supplied by the Widnes Foundry Company, carries the logo of the Victoria Pier Company. Restoration of the pier is far from complete, but in 2001 it was found that proposals for renovation were unlikely to gain support from the Heritage Lottery Fund.

Right: *Bangor Garth Pier suffered collision damage in 1914, with permanent repairs and improvements finally being made in 1921 when it was found that the central piles had settled 4.5 inches (11.4 cm). The pier closed in 1971 for safety reasons but was re-opened after restoration in 1988. The curved ironwork that forms seating along the sides of the pier is especially elegant.*

As pier and pavilion design required a combination of engineering and design skills, it was not surprising that many firms, including engineers, general contractors, ironfounders and prefabrication specialists, became involved in the process, along with architects whose practices happened to be based at resorts or in their catchment areas. The seaside became a profitable arena for the ironfounders, whose catalogues provided much of the

Great Yarmouth Britannia Pier's first permanent pavilion, a triumph of prefabrication, which opened in June 1902. The pier itself was the second on the site; the original 700 foot (213 metre) long Britannia Pier was built in 1857–8 but demolished after the 1899 summer season and replaced with a slightly longer pier, which opened in 1901. The second pier's designers were the Manchester brothers James and Arthur Mayoh, the contractors being their firm Mayoh & Haley.

Above: *Cleethorpes Pier was built in 1872–3 for the Cleethorpes Promenade Pier Company; its original length was 1200 feet (366 metres). In 1884 the pier was taken on by the Manchester, Sheffield & Lincolnshire Railway Company, which added a pavilion in 1888, but this was burnt down in 1903. The pier is now 335 feet (102 metres) long as a result of breaching in 1940 and demolition after the war.*

Left: *Cromer Pier (1900–1) stands directly below the Hotel de Paris, built (around existing properties) in 1895–6 by the Norwich architect George Skipper (1856–1948). He designed two other massive hotels at the little resort, the Grand (1890–1) and the Metropole (1893–4), but both were demolished in the 1970s.*

decorative pier furniture as well as the substructure. Boulton & Paul of Norwich, specialists in prefabricated timber-and-iron structures, illustrated a substantial pavilion as supplied to clients at Scheveningen, on the Dutch coast, in their 1898 catalogue and provided a larger version in 1902 for Great Yarmouth's Britannia Pier. It was probably the largest of the prefabricated pier buildings until it was destroyed by fire in 1909.

After the pier-construction boom of the 1860s and 1870s, when an average of two new piers per year appeared on the British coastline, pier development in the final quarter of the nineteenth century concentrated upon the pavilions, which were transformed from functional shelters to ornate palaces of entertainment. They were an eclectic bunch in architectural terms, ranging from Birch's stunning Indian Pavilion on Blackpool North Pier (1874, destroyed by fire in 1921) to

Head Wrightson were contractors for Cleethorpes Pier, erecting the iron piling, with its decorative capitals, and the unusual arched ironwork spans. The shore-end pavilion dates from 1905 and – although much altered – still retains its original roof structure.

Southsea's Clarence Pier was built in 1860–1, the pavilion being added in 1882; the stubby pier's width is greater than its length. The resort's second pier, South Parade, was constructed during 1875–8 by Head Wrightson and opened in 1879. A fire in 1904 resulted in a complete rebuilding, which took until 1908.

Clacton Pier's oval-plan seaward-end pavilion was built in 1893, the curving ironwork of its structure still visible following storm damage to later cladding. The pavilion was designed as a concert hall. A shore-end pavilion, the Ocean Theatre, was added in 1928, followed by the Crystal Casino in 1932 (demolished 1940).

Looking shoreward on Southend Pier, Britain's longest pier at 7080 feet (2158 metres) and nearly twice as long as Southport Pier, the second longest. The first pier at Southend opened in 1830 and was extended to become the longest pier in Europe by 1846. The present pier is based on the iron pier built in 1888–90 by James Brunlees.

All that remains of the winter-garden-style pavilion (1883–4) at Llandudno Pier is a series of ironwork columns still carrying their decorative spandrels and the iron stairs reaching to the sky. The pavilion, which was a 2000-seat theatre with a cast-iron veranda on the seaward side, was repaired during the 1980s but was empty from 1990; a 1994 fire brought about its demolition.

Blackpool Tower seen from beneath the Central Pier, with the North Pier in the distance. The Central (originally South) Pier was built in 1867–8 by the contractor Robert Laidlaw; its iron columns were placed at 60 foot (18 metre) intervals and linked by wrought-iron girders. The pier was always renowned for its popular entertainments.

Left: *Lowestoft's Claremont Pier was 600 feet (183 metres) long and used by steamers when it was built in 1902–3. The last steamer called in 1939 and the pier was derelict by 1948, although a new shore-end pavilion was built shortly afterwards. Restoration of the pavilion began in 1988, and a colourful (and successful) beach-hut promenade development was introduced just to the south in the late 1990s.*

Llandudno's shore-end iron-and-glass winter garden (1883–4, demolished after a fire in 1994) and the Arts and Crafts style cottage on Lowestoft's South Pier (1891, demolished late 1940s). During the 1890s and early 1900s a series of huge marine palaces, mostly oriental in style, was built on the piers of Sussex and Lancashire, the most spectacular being the 'Taj Mahal of the North' added to Morecambe Central Pier in 1897; it was the work of specialist seaside

The pavilion known as the 'Taj Mahal of the North' was added to Morecambe Central Pier in 1897. It was built by the Morecambe Pier and Pavilion Company to enable their pier – which opened in 1869 and had few entertainment facilities – to compete with the town's new West End Pier, which was not only much longer but came equipped with a seaward-end pavilion.

The 1800 foot (549 metre) long Morecambe West End Pier (1893–6) was built by the contractors Mayoh & Haley with prefabricated iron parts supplied by the Widnes Foundry Company. The large domed pavilion, an integral part of the original design, was destroyed by fire in 1917; the pier itself was demolished in 1978.

architects Mangnall & Littlewoods of Manchester. The most unusual surviving pavilion of this era was designed by Great Yarmouth's borough surveyor J. W. 'Concrete' Cockrill for the town's Wellington Pier in 1903. Its idiosyncratic clean-cut modern style is reminiscent of contemporary exhibition buildings, although its decorative finials were lost when the (now rather neglected) pavilion was re-clad in the 1950s.

Although nine piers were completed during the first decade of the twentieth century, eight of these opened between 1901 and 1905, with Weston-super-Mare Grand Pier (1904), complete with an integral 2000-seat pavilion, being the most impressive. Fleetwood Pier (1910), only 492 feet (150 metres) long, and Burnham-on-Sea Pier (1911), an even shorter structure built from concrete, were the

In Great Yarmouth, Wellington Pier's pavilion, originally clad in metallic Uralite, is still a striking and unusual structure despite losing many of its decorative elements in the 1950s. Its architect, John William Cockrill (1849–1924), was particularly interested in building with ceramics and concrete, even attempting to persuade property owners to concrete over Yarmouth's narrow lanes (known as rows). This earned him the nickname 'Concrete' Cockrill.

The massive pavilion on Weston-super-Mare's Grand Pier dates from 1932–3; it replaced the original pavilion, which burned down in 1930. The pier itself is an iron-piled structure erected by the contractors Mayoh & Haley; its framework of lattice girders is carried by cast-iron columns grouped in tens. A canopied shelter runs the entire length of the neck, and the pier is listed Grade II.

last pleasure piers to be built before the Second World War. Pavilions, however, continued to be added to existing piers. Most were small domed or barrel-vaulted halls with little decoration, or even prefabricated structures, but there were exceptions: a Moorish-style pavilion was added to the pier at St Anne's-on-the-Sea in 1904, while Brighton West Pier's Concert Hall (1914–16), its delicate ironwork engineered by Noel Ridley, could seat up to 1400 people. This pavilion was the home of the West Pier Orchestra, although its floor could be cleared to host other activities such as roller-skating or ballroom dancing. The range of activities available on piers, which by the early twentieth century included music and shows (everything from German bands and minstrels to pantomime, opera and music hall), aquatic performances and slot-machines, expanded still further after the First World War to take in fairground rides, cinemas, vast dance halls, sunbathing and sophisticated amusement

Brighton West Pier's first great pavilion (right) was constructed at its seaward end in 1893 by the engineer Peregrine Birch (1845–96), son of Eugenius Birch, builder of the West Pier. Brighton Corporation forced the removal of a large dome from the younger Birch's pavilion design. The Concert Hall (left) was designed by Clayton & Black and added in 1914–16, its completion being delayed by the war. The interior featured coloured designs stencilled on the walls.

Looking south along the beach at Great Yarmouth towards the Wellington Pier and Winter Gardens. The Winter Gardens, which opened at Torquay in 1881, was designed by local architects John Watson and William Harvey, while the ironwork was supplied by the Crescent Iron Works of Willenhall. It was never successful and was bought by Great Yarmouth council in 1903 for £1300, dismantled, shipped round the coast and re-erected. It enjoyed happier times in Yarmouth, which prospered in the Edwardian era.

arcades. The piers of the 1920s and 1930s were just as popular as those of the late nineteenth century.

Funding for pier improvements increasingly came from local councils, anxious to promote tourism. Great Yarmouth council bought the town's Wellington Pier in 1900, funding its reconstruction and the addition of a new pavilion as well as the transport of a winter garden from Torquay to add to the

Morecambe Central Pier's exotic Indian-style pavilion was destroyed by fire in 1933; a 2000-seat replacement pavilion-cum-ballroom was put up in 1935–6. The pier closed in 1986 after partial collapse of its decking; then the pavilion was burnt down in 1991; complete demolition of the pier followed in 1992. It had at least outlasted the rival West End Pier by well over a decade.

A train on Southend Pier approaches one of the pier's shelters. The pier, built from rolled steel joists (horizontal members) on cast-iron screw piles, is now served by lightweight electric trains running on 3 foot (1 metre) gauge tracks, the single-track railway being completely separated from the footway.

entertainment complex. In Worthing, the corporation bought the pier in 1920, repaired it and provided it with a shore-end concert pavilion in 1925–6 at a total cost of almost £60,000; the borough architect designed its highly glazed modernist, metal-clad seaward-end pavilion in 1935. Financial difficulties suffered by pier-building companies often necessitated the involvement of local councils; continued deterioration of the fabric of Victorian piers and pavilions meant resorts were left without their most attractive amenities.

Several pavilions fell victim to fire, notably Birch's Indian Pavilion at Blackpool North Pier (1921) and the 'Taj Mahal of the North' on Morecambe Central Pier (1933). Great Yarmouth's unlucky Britannia Pier suffered a series of fires, beginning in 1909 with the destruction of the first pavilion (1902). The second pavilion (1910) was burnt out in 1914 and the Floral Hall Ballroom (1928) followed in 1932. During a fire in 1954 its replacement, the Grand Ballroom (1933), perished, as did the third pavilion in 1914. By 1939, although 101 pleasure piers had been erected, twenty-five had already been demolished or rebuilt, leaving seventy-six to face the rigours of the coming war.

Modern piers

The Second World War was especially disastrous for piers on the south and east coasts, many of which were breached to prevent their use as landing-stages; floating mines and enemy action also caused damage. A worse fate befell Minehead Pier, built in 1900–1, which was completely demolished in 1940 to provide nearby gun batteries with a clear line of sight; in the same year, Deal Pier was taken down following a ship collision. After the end of the war the remaining piers, tattered and broken, were returned to their owners. Some financial compensation was available, but eight piers were eventually lost during the 1950s, mainly as an indirect result of combined wartime and storm damage, including St Leonards Palace Pier (1888–91) with its fine oriental-flavour pavilion. Here, breaching, bombs and fire had severely affected the pier's structure, and demolition followed further gale damage in 1951.

Many piers were rapidly reconstructed and were able to reopen soon after the war ended, although structural damage to others resulted in considerable cuts in their deck length. At Deal a completely new pier was built in 1954–7 with reinforced concrete, a material that had been used for pier repairs since the mid 1920s. The design, by Sir William Halcrow and Partners, comprised concrete decking above slightly inclined piles; although traditional in plan, the effect was wholly modern and very unlike the eccentric, colourful architecture of the 1951 Festival of Britain, which included the Seaside Pavilion, part of which was intended to resemble a pier. Deal was the sole new pleasure pier to be erected in the second half of the twentieth century, although the continued popularity of the

The no-frills concrete structure of Deal Pier, which reaches out 1026 feet (313 metres) into the English Channel. The pier, which cost £250,000 and was opened by the Duke of Edinburgh in 1957, is basically a promenade pier with fishing facilities.

The 'Seaside Pavilion' at the Festival of Britain, drawn by the artist Bryan de Grineau and reproduced in the 'Illustrated London News' of 26th May 1951. Visitors on the left can be seen looking out over the Thames from a pier-like structure supported by lattice girders. The pavilion, designed by Eric Brown and Peter Chamberlin, also included rock making, saucy postcards and donkey rides, all intended to represent the 'essentially English' seaside.

British seaside holiday during the 1950s led to the construction of some new pavilions whose design was rather different from that of their predecessors, for instance Lowestoft South Pier's glass-walled stairwell topped by an observation platform.

During the 1960s and 1970s, before appreciation of the delights of Victorian architecture became widespread, another eight piers were demolished. Only fifty-nine piers remained when the National Piers Society was founded in 1979, its aims being to promote and sustain interest in the preservation and enjoyment of seaside piers. The destruction of Redcar Pier in 1981 – sold off for just £250 – seemed to signal a decade in which pier restoration was greeted with enthusiasm by the public and became acceptable to funding bodies. Bangor Garth Pier, closed in 1971 and threatened with demolition, was restored during 1982–8,

Southend Pier was threatened with demolition even before the fire in July 1976 that severely damaged the pierhead; although rebuilding began in 1984, the burnt timbers still remain. The new lifeboat station, at the far end of the pier, is seen under construction during 2001.

Above left: *After its closure in 1971, Bangor Garth Pier was saved from demolition by Bangor City Council, which bought the pier for one penny in 1975. Raising funds for its renovation then took a further seven years, and restoration work costing £3.5 million was carried out during 1982–8. However, high running costs forced the council to put the pier up for sale in 1999.*

Above right: *Cromer Pier opened in 1901, replacing two earlier jetties, both of which had been washed away by heavy seas. The seaward end was extended in 1905 when the pavilion was built; this still functions as a traditional summer-season variety theatre.*

while Clevedon Pier (now Grade I listed) reopened in 1989 after almost two decades of fund-raising and rebuilding, although its full restoration was completed only in 1998, when Swanage and Penarth Piers were also reopened following renovation. Financial support for the restoration of piers from the Heritage Lottery Fund during the late 1990s, including funding for Brighton West Pier, Britain's only other Grade I listed pier, followed three further demolitions (Morecambe Central, Ventnor and Shanklin), leaving just fifty-five pleasure piers around the British coast in 2000.

Refurbishment of Hastings Pier continued during 2001, with part of the pier being opened for the summer season. From a distance the new white buildings appear reasonably attractive, although one pavilion has been remodelled as a small shopping mall with shops that are completely cut off from views of the sea, rather negating the point of their situation on the pier.

21

The Hastings Pier frontage was rebuilt between the First and Second World Wars using Ceramic Marble, a material produced by Carters of Poole that was highly resistant to seawater. The pier entrance, which was altered again during the 1960s, is opposite the White Rock Pavilion, a theatre built in 1927 and decorated with colourful pictorial relief roundels in Doulton's Polychrome Stoneware designed by the sculptor Gilbert Bayes.

Brighton West Pier in 2001, with only the tiniest of connecting bridges between pavilions and shore. The pier, which has been under threat of demolition since the 1970s, was purchased by the Brighton West Pier Trust in 1984. Storm damage in the late 1980s cut the link between pavilions and shore, but following a grant from the Heritage Lottery Fund in 1996 a steel walkway was built to cross the gap.

Brighton Palace Pier was built in 1891–9 at a cost of £137,000; its length was 1760 feet (536 metres). A 1500-seat theatre opened at the seaward end in 1901 and a pavilion was added to the central section in 1910; the theatre was dismantled in 1986. The pier's delicate ironwork was part of the original oriental-style design by R. St George Moore, who was also responsible for the pier at St Leonards, built in 1888–91 and demolished in 1951.

The Noble Organisation acquired Brighton Palace Pier (now known as Brighton Pier) in 1984 from the Brighton Marine Palace and Pier Company, its original owners. This popular pier is now home to a range of entertainment facilities including a funfair with a helter-skelter, something of a landmark, near the end of the pier.

Saltburn Pier opened in 1869, when its length was 1500 feet (457 metres); a long history of storm damage and other problems including a ship collision (1924) ended with restoration during the 1970s and reopening in 1978 at a length of a mere 681 feet (208 metres). The pier is now the last remaining pleasure pier on the Yorkshire coast, and further restoration of the deck took place in 2000–1.

Preservation and restoration of these enjoyable structures, which blur the distinction between engineering and architecture, continues into the twenty-first century. Their apparently timeless appeal has ensured that most of the remaining piers are supported by pier-preservation trusts or individual pier owners who continue to take on these difficult but rewarding structures. Saltburn Pier was restored in 2000–1, with support from Redcar and Cleveland Borough Council, the Heritage Lottery Fund and the European Regional Development Fund as well as the Friends of Saltburn Pier. In Suffolk, the stubby 150 foot (46 metre) remnant of Southwold Pier, bought by Chris and Helen Iredale in 1987, has benefited from an almost complete reconstruction, taking its length to 620 feet (189 metres) and including a series of pavilions. New Southwold Pier, which may be regarded as the first new pier of the century, was opened by the Duke of Gloucester on 3rd July 2001 in front of a crowd of around 8000; the owners hoped for 20,000 or more promenaders in their first summer season.

Construction of New Southwold Pier and its pavilions taking place during 2000. A T-section pierhead is also to be built, enabling passenger vessels to call at the pier, as the 'Belle' steamers did before the original pierhead was destroyed by storms in 1934. The pier, which reopened in 2001, now offers visitors a pier history exhibition, shop, pub and diner as well as an amusement arcade.

Coastal steamer traffic declined between the First and Second World Wars, but its architectural legacy includes Wemyss Bay station, where day-tripping Glaswegians left the train to catch the Rothesay ferry; the adjacent pierhead is reached via a curving, covered wooden walkway. The station, which was rebuilt by the Caledonian Railway in 1903, was designed by the architect James Miller and the engineer Donald Mathieson, who both worked on the extension of Glasgow Central Station (1899–1905).

The people's pleasure palaces

Such was the popularity of the seaside holiday in the late nineteenth and early twentieth centuries that huge entertainment complexes became necessary to cater for the crowds. These massive leisure facilities, of which Blackpool Tower is the best known, were funded by an array of financial backers, from large industrial concerns to a multitude of small investors, and were largely located within reach of the cities of north-west England. English visitors and sometimes English finance helped the entertainment business flourish in Wales and the Isle of Man, but the English holiday-making masses never reached Scotland. There, resort development was also restricted by climate, and only a few resorts, of which Rothesay was the most

Tynemouth Winter Garden was built above the resort's Long Sands in 1876–8, its designers being the London partnership of John Norton, architect to the Crystal Palace Company estate, and Philip Masey. The spectacular structure included a basement aquarium and a winter garden over 200 feet (61 metres) long. The building was never a financial success, and, despite being Britain's sole remaining brick-façaded winter garden, was not a listed building. It burned down on 10th February 1996.

The tiered structure of Torquay (later Great Yarmouth) Winter Gardens was more like that of a horticultural winter garden than the great seaside conservatories at Southport and Bournemouth. The Torquay glasshouse had no curved panes and little decoration apart from floral motifs on the ironwork. It was 170 feet (52 metres) long and measured 83 feet (25 metres) to the top of the lantern.

significant, could support entertainment buildings approaching the English ones in scale.

The idea of these pleasure palaces originated with the smaller and earlier winter gardens and assembly rooms. The success of the Crystal Palace, initially at the Great Exhibition of 1851 and then at its Sydenham site from 1854, inspired the construction of many smaller iron-and-glass winter gardens at the resorts. However, it was the series of orchestral concerts held at the Crystal Palace, drawing audiences as large as 80,000, that made fashionable the combination of music and large-scale winter garden. Three massive seaside winter gardens were built: at Southport in 1874, where the Manchester architects Maxwell & Tuke built what was advertised as 'the largest conservatory in England'; at Bournemouth (1875–6); and at Torquay (1878–81). The last is the only survivor, having been sold to Great Yarmouth council in 1903, dismantled, shipped round the coast and re-erected at the town's Wellington Pier; the Southport and Bournemouth winter gardens were demolished during the 1930s. Along with the Crystal Palace, Torquay's winter garden is one of the few theoretically demountable buildings actually to be moved and rebuilt.

Another international exhibition, the 1889 Paris Exposition Universelle, provided the next stimulus for seaside building. Its iconic Eiffel Tower, then the tallest building in the world at 984 feet (300 metres), was visited by nearly two million people during the exhibition and rapidly recouped its cost. The idea of a tower as a profitable attraction quickly took root at the British seaside, where the tower could be seen as a vertical version of the pier, combining a view with other entertainments. Patents for towers in a wide variety of shapes, heights and construction systems

The exterior of Blackpool Tower Buildings, overlooking the promenade, was far from decorative, its combination of red brick and elaborate terracotta dressings making the structure appear more like a department store than an entertainment centre. Its original design did indeed include two shopping arcades, although these were never built.

Blackpool Tower's Grand Pavilion was converted to the Tower Ballroom by Frank Matcham in 1898–9. The style was French Renaissance and the effect overwhelming; the ballroom, said to be one of the three finest in Britain, was an instant success. Matcham's important interior was burnt out during 1956 but was reinstated to exactly the original design, reopening in 1958.

Right: *In contrast to its forbidding exterior, Blackpool Tower's interior is highly decorative. This low-relief Burmantofts faience panel with rich turquoise glaze is one of a series lining the main staircase; the designer was E. Caldwell Spruce, a principal modeller at the Leeds firm. Other panels show fishes, cherubs and children.*

were granted, although most of these were never built. The first attempt, at Wembley (where the tower was intended to be the centre of a vast outdoor amusement park), reached the height of 155 feet (47 metres) before funding ran out; this failure appeared to deter potential investors in seaside towers.

A tower for Blackpool was originally proposed by the Standard Contract and Debenture Corporation, a property company with Isle of Man and London connections, although the more locally based Blackpool Tower Company took over the site in 1891. A design competition for the tower was won by Maxwell & Tuke and construction began in mid 1891; R. J. G. Read of Westminster was the consulting engineer. The 500 foot (152 metre) tower itself was built by the end of 1893 but the complex of buildings below was only

Blackpool Winter Gardens was given a dramatic new appearance from 1929, when its façade was clad in sparkling white faience supplied by Shaws of Darwen. The order, which included blue and yellow decorative elements such as the lettering emphasising the archway, was worth £22,638 to the firm.

New Brighton Tower, completed in 1898, was an octagonal complex of buildings in red Ruabon brick with terracotta and stone dressings; the octagonal tower rose from its centre. The unsuccessful tower met the same fate as the town's pier, which was closed in 1972 and demolished in 1977.

finished after the Tower opened to the public on Whit Monday 1894. Early customers eventually experienced an accumulation of delights previously unknown at any single British entertainment venue: an aquarium and menagerie, a circus cunningly fitted into the area between the four legs of the tower, the lift ride to the top of the tower, the Grand Pavilion (later the Tower Ballroom) and a range of shops, bars and cafés, all in a building that was imposing without and highly decorative within. Colourful high-relief Burmantofts ceramic panels brightened the corridors while the circus was tricked out with glittering, Moorish-style ceramic ornament.

The Tower was an undoubted success, much to the consternation of other Blackpool entertainment promoters, especially those of the nearby Winter Gardens. The Winter Gardens began life in 1878 as an elegant, apsidal-ended theatre with a glass-roofed promenade running around its base. To this was added in 1889 an opera house by the leading theatre designer Frank Matcham, and then, stung by competition from the Tower, the ornate Empress Ballroom (by the architects Mangnall & Littlewoods) in 1896. The Tower and Winter Gardens continued trying to outdo each other until the turn of the century; they were joined by a third pleasure palace, only a few streets away, when the Alhambra opened in 1899. This massive building, designed by the architects Wylson & Long, contained three huge performance spaces: a 2000-seat circus, a ballroom for 3000 dancers and a 3000-seat theatre. Even for Blackpool this was a palace too far and the receivers were brought in during 1902. The Blackpool Tower Company purchased the Alhambra in the following year, called in Frank Matcham to redesign its interior and opened it in 1904 as the Palace. In 1914 an underground passage was created joining Tower and Palace so that pleasure-seekers could avoid the promenade gales; however, the Palace was demolished in 1961.

At New Brighton, Wirral, another and taller tower was under construction in 1897, again designed by Maxwell & Tuke. The New Brighton Tower and Recreation Company initially found capital easy to come by as a result of Blackpool Tower's success and went for an altogether grander complex: the eight legs of the 576 foot (176 metre) high tower enclosed the 3000-seat Grand Tower Theatre, the largest theatre in the country outside London; there was also a

Morecambe Tower, which stood at the north-eastern end of the town's seafront from 1899, was bought by the Morecambe Tower and Estates Company in 1909 after its original owners had run into financial problems. More of the tower's base buildings were then completed, including the pavilion (1909) and a ballroom (1911), and the enterprise was marketed as a fairground (an early theme park), but with little success. The tower itself did not survive the 1914–18 war, but it was 1961 before the remaining base buildings were demolished.

ballroom with capacity for 2000, a winter garden and other entertainments. The landmark building opened in 1898 but could never draw the enormous number of visitors necessary for profitability; the tower itself was demolished in 1919–21 and the remaining buildings burned out in 1969. Even shorter-lived was Morecambe's attempt at a tower-cum-pleasure palace, the skeleton of which was built in 1899–1900. The Morecambe Tower, with all its base buildings and pleasure grounds, was intended to provide a

The architects of Southend Kursaal, seen here in 2001, were George Sherrin and John Clarke of London. The much-altered interior originally included a circus and a theatre, actually a flat-floored music hall that could double as a ballroom. The Burton brewing company Samuel Allsopp's had taken over the building by 1903, and later developments concentrated on the amusement park rather than the Kursaal itself.

The Queen's Palace at Rhyl, seen here around 1905. It was lost to fire in 1907, although another domed entertainment centre, the Rhyl Pavilion, was erected near the seafront in 1908. Rhyl Pier, built by James Brunlees in 1867, was demolished in 1973.

The theatre at the Queen's Palace, Rhyl, which was designed by C. J. Richardson and built in 1902. The theatre alone was rebuilt after the 1907 fire, without the domed pavilion, and is still extant, although its auditorium was reconstructed in 1928.

wholly oriental environment, including a bazaar and ethnic stallholders, its visitors being carried to the top of the cone-shaped tower by a spiral tramway. The venture did not prosper and the tower itself was dismantled during the First World War.

Away from the north-west, Southend Kursaal (whose guidebook compared its attractions to Blackpool's Winter Gardens) was rather more successful, its 1898 building being topped by a dome and including a range of attractions as well as ninety shops; the building, although much altered, still stands. The last of the pleasure domes, the Queen's Palace, was built at Rhyl in 1902 to cater mainly for Lancashire trippers. Under its great glazed dome was a ballroom with sprung parquet floor on which 2000 couples could dance, a theatre, a winter garden, forty shops, a zoo, a waxworks, a native village and – beneath the ballroom – an imitation Venice with real canals. The Queen's Palace was destroyed by fire in 1907, leaving only Blackpool's Tower and Winter Gardens and the much-changed Southend Kursaal to remind us of this era of mass entertainment and massive buildings.

The white dome of Spanish City (1911) originally marked the presence on Whitley Bay seafront of a leisure complex including a large flat-floored theatre. This unusual pleasure dome, 50 feet (15 metres) in diameter, was constructed using the Hennebique reinforced concrete system (first used in Britain in 1897); even in the 1970s the Spanish City dome was still thought to be second in size only to that of St Paul's Cathedral. The architects were the Newcastle upon Tyne practice of James Thoburn Cackett and Robert Burns Dick.

Even as some of the largest indoor entertainment complexes were still being built, outdoor entertainment on a similar scale, in the form of the amusement park, was becoming popular. At Blackpool the beginnings of a fairground appeared on the South Shore in the early 1890s, with larger rides being introduced from 1904. The fairground, which was known as the Pleasure Beach from 1906, had become the biggest and most modern amusement park in Britain by 1909; its owners formed the private company Blackpool Pleasure Beach Limited in 1910. Along with the largest of the seaside piers and their varied facilities, the amusement parks at Blackpool and elsewhere became outdoor pleasure palaces, all-purpose leisure environments that were the forerunners of modern theme parks.

The first British example of a Warwick's Revolving Tower (left) was built on Great Yarmouth seafront, just north of Britannia Pier, in 1897; the first ever had been put up in Atlantic City, New Jersey. The Yarmouth tower continued in operation until 1939, although the revolving mechanism had ceased to function after the First World War; it was demolished in 1941.

Entertainment ashore

When seaside resorts were simply watering places, leisure at the seaside was an elegant affair of quiet enjoyment, simple contemplation of the view from a promenade shelter, but with the crowds came the need for novelty attractions that were also profitable. The view from the promenade was free but could be commercialised by means of enticing the customer to pay for its enhancement. A series of observation towers known as Warwick's Revolving Towers was built by the London engineer Thomas Warwick at Great Yarmouth, Cleethorpes, Morecambe, Douglas, Scarborough and Southend during 1897–1902. These inelegant latticework structures, about 150 feet (46 metres) high, were hexagonal in cross-section and carried a platform on which a carriage revolved. Thus up to 200 passengers at a time

Saltburn's inclined tramway (foreground) opened in 1884. It was Britain's third cliff lift, the first two having been built at Scarborough in 1875 and 1881. The water-balanced tramway replaced a vertical hoist that had carried passengers between cliff and beach during 1870–83; the hoist was designed by John Anderson, who was also engineer for Saltburn Pier.

Blackpool Illuminations in 1996. The Tower has featured in the Illuminations since 1925, when its lighting carried the message 'Wonderland of the World'. More than ten thousand bulbs are required to light the tower.

were simultaneously lifted and rotated to enjoy a panoramic view; the power was provided by a steam engine that generated electricity. Cliff railways also offered the combination of view and transportation. The first seaside cliff railway was built at Scarborough's South Cliff in 1875 and others followed, for instance at Saltburn (1884) and Hastings East Cliff (1903); the station buildings, their design reminiscent of small tollhouses, often form an attractive part of the townscape.

The view could also be transformed by the use of electric lighting. Blackpool Illuminations began life in 1897 when the first illuminated trams trundled along the promenade to mark Queen Victoria's Diamond Jubilee. The combination of coloured lights and movement was instantly successful, and the concept was expanded from 1912 to include strings of electric lights hung from lamp-posts and across arches. Blackpool Casino, built at the entrance to the Pleasure Beach in 1913, was one of several contemporary seaside buildings to be defined by their lavish use of electric lighting. This 700-seat cinema was constructed from reinforced concrete in the style of an Indian palace; at night, the fantastic domed Casino glittered at the end of the promenade, drawing holiday-makers toward the Pleasure Beach. Great Yarmouth's Gem Cinema, built in 1908 by the local architect Arthur S. Hewitt, was known as 'The Palace of 5000 Lights', although this effect was actually achieved by fixing 1500 electric light bulbs to its ornate buff terracotta

The Casino, a cinema, was built at Blackpool Pleasure Beach in 1913. From 1910 the Pleasure Beach began to provide indoor entertainments, the first being the Naval Spectatorium, a massive circular building in which American Civil War naval battles were replayed using lighting and mechanical devices.

Above: *Great Yarmouth's Gem (later Windmill) Cinema, which opened in 1908, was probably England's first purpose-built seaside cinema, although the flat-floored hall could also be used for variety performances.*

Right: *Great Yarmouth's Empire Cinema opened in July 1911; it was built for the Yarmouth and Gorleston Investment and Building Company and was designed by the architect Arthur S. Hewitt, who was also responsible for the nearby Gem Cinema. Its Edwardian baroque façade, a triumphal arch framed by giant, fluted Ionic columns, was clad in Burmantofts vitreous terracotta. The interior was in sumptuous Louis XIV style.*

Below left: *Great Yarmouth Hippodrome, an unusual purpose-built circus seating 2500, was designed by the architect Ralph Scott Cockrill (born 1879) and opened in 1903. Three stained-glass windows depicting chariot racing originally graced the entrance façade.*

Left: *The manufacturer of Great Yarmouth Hippodrome's delicate Art Nouveau terracotta is unknown, but may perhaps be Doulton of Lambeth. The circus had a double-roof structure and 2 foot (0.6 metre) thick concrete walls to provide adequate noise insulation.*

33

Left: *The architect of Great Yarmouth Hippodrome, Ralph Scott Cockrill, had at least two younger brothers who became architects, while his architect father – the borough surveyor J. W. Cockrill – was also the inventor of the Cockrill wall facing tile patented with Doulton of Lambeth in 1893. Ralph Scott Cockrill's grandfather was William Cockrill, a builder and developer from nearby Gorleston-on-Sea.*

Below: *At the centre of Torquay's seafront is the Pavilion (1911–12). Its original 1896 design was by Edward Rogers, but construction was completed after his death by Torbay's borough engineer Henry Augustus Garrett. The steel-framed building is clad in white Doulton Carraraware with pale-green banded decoration. The Pavilion was saved from the threat of demolition in the early 1970s and remodelled in 1986–7 as a shopping arcade; its 1939 extension was removed, and the eastern façade reconstructed in specially manufactured faience blocks.*

façade. There were also several water parks whose exotic structures were illuminated by night, as at Scarborough's Northstead Manor Gardens or Great Yarmouth's Waterways, where the boats were decorated fairground-style and daytime colour was provided by mass flower-bedding. Colour, lights and water also came together at the circus, where the grand finale at technically sophisticated venues like Great Yarmouth's Hippodrome, built in 1903, was the water spectacle, initially performed in Rome's Colosseum; at the Blackpool Tower circus it took only two minutes for 40,000 gallons of water to flood the ring and transform the arena.

Up to the mid 1920s, traditional seaside design changed little,

Rothesay Winter Gardens originated around 1895 as an octagonal bandstand on the Esplanade. In 1923–4 a circular domed structure, prefabricated at Walter MacFarlane's Saracen Foundry, was added to cover the seating area, while the bandstand was also roofed over and an entrance was created between twin pagoda-style square towers. The building is now the Isle of Bute Discovery Centre.

Right: *The domed interior of Rothesay Winter Gardens, showing some of the sixteen steel radial supporting ribs that centre on a pendulous faience light fitting. There is much Art Nouveau cast-iron ornament on the building's exterior (where a walkway circles the dome), and even the iron ventilation grilles are highly decorative.*

retaining its emphasis on prefabrication and cosy vernacular but sometimes idiosyncratic structures. Winter gardens in various formats remained popular: Walter MacFarlane's Saracen Foundry in Glasgow supplied prefabricated components for the not dissimilar pair Ryde Pavilion (1923–6) and the domed Rothesay Winter Gardens (1923–4), both of which still survive, but the seaside was soon to become the most receptive location in Britain for architecture of the Modern Movement.

The first significant Modernist structure to appear at the seaside was the Royal Corinthian Yacht Club, designed by the architect Joseph Emberton and built at Burnham-on-Crouch in 1931. Its long, low clean white lines and extravagant horizontal glazing brought to mind sunshine, elegance and a streamlined brand of leisure rather different from that associated with Victorian decorative excesses. Other fine examples of international Modernism followed, including the Midland Hotel, Morecambe (1932–3), by the architect Oliver Hill,

The Royal Corinthian Yacht Club, built at Burnham-on-Crouch in 1931. The steel-framed structure stands on a concrete platform supported by stilts in the water, its long horizontal windows providing commanding views across the estuary. It was a dramatic and initially controversial replacement for the Yacht Club's previous premises, a small weather-boarded clubhouse.

built for the London Midland & Scottish Railway and a descendant of many great resort hotels, the grandest of which was Scarborough's Grand Hotel (1863–7). The Midland was rather Art Deco in feel and featured artworks by Eric Gill and Eric Ravilious as well as rugs by Marion Dorn. The much-photographed De La Warr Pavilion, Bexhill (1933–6), was designed by architects Erich Mendelsohn and Serge Chermayeff and offered the somewhat surprised residents of the little seaside town a repertory theatre, tea

dances, a restaurant and a reading room. Afternoon tea may still be taken on its serene sunlit balconies.

Less well known examples of seaside Modernism include Ventnor Winter Gardens (1935), Rothesay Pavilion (1938) and even the now demolished Derby Baths (1938–9) in Blackpool, a daring yellow and lime-green ceramic box whose walls were cut through by porthole windows and decorated with fishy roundels; the faience was by Shaws of Darwen while the design was by the borough surveyor, John C. Robinson. Also in Blackpool, the Pleasure Beach Casino was demolished in 1937 to be succeeded in 1939 by Joseph Emberton's circular white Casino, its dizzy spiral stair tower suggestive of delights within the amusement park.

Earl De La Warr, a socialist, was Mayor of Bexhill during 1932–4 and the little town's major landowner. In 1933 he organised an architectural competition for a Modernist public building, attracting 230 entries, and put up £100,000 to see construction of the welded steel-framed De La Warr Pavilion completed in 1936. Controversy over its design and philosophy brought his period as Mayor to an end.

The broad half-drum of the café at Rothesay Pavilion (1938) appears to have taken at least some inspiration from Bexhill's Pavilion; the Rothesay design was a competition-winning entry from the architect James Carrick. Inside is a rectangular auditorium and a grand Art Deco staircase. Although the stylish building is still very much in use, the ground level glazing has been (temporarily) defaced by advertisements.

Blackpool Casino, built at the Pleasure Beach in 1939, rests on a base of reinforced concrete above the sandy subsoil. The Trussed Concrete and Steel Company (Truscon) designed the Casino's reinforced-concrete structure; the thin pre-cast concrete facing panels were made in Holland, then delivered to Blackpool by barge. Inside the Casino was a series of restaurants, everything from First Class to Soda Fountain, while its banqueting room could seat 1000 people.

Above: The Jubilee Pool, Penzance (1935), is superbly situated on a promontory, with views towards St Michael's Mount and Newlyn. The tidal pool was designed by the borough engineer F. Latham and incorporated stepped terraces for sunbathing as well as floodlights to allow for swimming after sunset.

Left: Saltdean Lido (1938), designed by R. W. H. Jones, was the first British lido to become a listed building. Its central two-storey pavilion contained a café and sunbathing terraces, while changing rooms occupied the wings; a diving board originally stood opposite the café. A tunnel beneath the coast road links the lido to the beach.

Paignton Festival Hall, built on the town's Esplanade in 1962 and designed by the architect C. F. J. Thurley. Just to the north is Paignton Pier, built in 1878–9 to the design of local architect George Soudon Bridgman (1839–1925), which became a regular port of call for pleasure steamers crossing Torbay; it was reopened in 1995 after restoration work.

Right: *Torquay's Princess Theatre at dusk, seen from Princess Pier. The theatre, in style a Scandinavian variation on the Festival Hall theme, was built in the 1960s, along with the pier's central shelter (1965), its angled glazing being a typical 1960s affectation. The pier itself dates from 1890, although most of its later steel substructure was replaced during 1978–9 as it had become corroded.*

Emberton worked at the Pleasure Beach from 1933, building many rides for the owners, who wished for a 'unified modern design'; fun and Modernism were securely joined in the public mind. Perhaps the purest expression of seaside modern style were the lidos, in particular the languid curves of the Jubilee Pool in Penzance (1935), Saltdean Lido (1938) and the massive New Brighton Lido (1934), said to have the largest outdoor pool in Europe but demolished in 1990.

A trickle of good seaside buildings continued to be constructed into the 1950s and early 1960s, for instance Dunoon Queen's Hall (1955–8) by the Glasgow architect Ninian Johnston, who used a jolly

Lowestoft's East Point Pavilion (1997), a café and information centre, with the Royal Norfolk and Suffolk Yacht Club (1902–3) in the background. The Norwich architect George Skipper produced an L-shaped plan to utilise the Yacht Club's difficult triangular site; the building is topped by a fully glazed observation room under a copper dome.

The Landmark Theatre was built on Ilfracombe's seafront in 1997 to replace the Victoria Pavilion Theatre (1925), which had been damaged by a storm in 1990. The designers of the striking conical structures, their forms based on traditional glass-making kilns or cones, were Tim Ronalds Architects. Over 300,000 bricks, made from German clay fired in Belgium, were laid during construction of the Landmark, which provides two main performance spaces, one of them flat-floored.

variety of materials in this Festival of Britain inspired replacement for an Edwardian pavilion opposite the pier, and Paignton's Festival Hall (1962), a colourful presence on the promenade. However, the decline in seaside holiday-making after the 1960s produced an increase in buildings designed to attract a new type of visitor to the resorts. Sadly, the results – including several conference centres and leisure centres, as at Cleethorpes, where the Leisure Centre (1983) dominates the southern end of the beach – were often nondescript, bulky and architecturally undistinguished.

Latterly, new thinking about the changed role of the resorts has brought high-quality architecture back to the seaside, where it belongs. During the 1990s the Tate St Ives (1993) led a renewal of interest in the Cornish resort and its artists, while Lowestoft's East Point Pavilion (an updated winter garden opened in 1997) at last went some way to capitalising on the resort's easterly location. More exciting is the National Trust's hi-tech White Cliffs Centre (2000), Dover, which brilliantly exploits the

This fishy artwork above the flood barrier on Southport's Marine Drive formed part of the 1998 programme of improvements to the town's sea defences, along with a pair of monstrous fish sculpted in the concrete of the sea wall adjacent to the pier.

Looking south along the South Promenade Improvement Scheme at Bridlington, a creative collaboration between the artist Bruce McLean and the architects Bauman Lyons. The design included buildings (a café-bar, chalets, toilets, office, deckchair store), landscaping (built-in terrazzo street furniture, a paddling pool) and public art knitted together by the 'Nautical Mile', a mile-long terrazzo pavement sculpture with inset text by Mel Gooding.

ever-changing view of shipping below, and Ilfracombe's Landmark Theatre (1997), where the nautical style of the 1950s has been surprisingly reinvented and renewed. Artworks have been integrated into promenade developments at Southport (partly driven by the need to improve sea defences) and in the South Promenade Improvement Scheme at Bridlington, which opened in 1998 and won a RIBA award. Here, collaboration between artists and architects resulted in a promenade environment so successful that visitor numbers increased by 20 per cent. Another art-based initiative, the proposal to build a Turner Centre at Margate, was already partially funded in 2001; architects Snøhetta and Spence won its design competition with a landmark structure in the form of a boat's hull clad in English oak, a fitting symbol for seaside regeneration.

The competition-winning design by the Anglo-Norwegian architectural collaboration of Snøhetta and Spence for the Turner Centre in Margate. The gallery, a powerful oak-clad form resembling an upturned boat keel, rises from the seabed while the smaller element, a bookshop and restaurant, sits on the pier from which it is approached. The visual arts centre is intended to be an important element in the regeneration of the resort; over half of its projected £7 million cost had been raised by late 2001.

Living and working at the seaside

Far from being a mundane backdrop for exotic entertainment architecture, some seaside homes – from the delightfully Gothick *cottages ornés* of Regency Sidmouth to holiday villages such as Thorpeness or Portmeirion – are equally inventive. Although the flat-roofed Modern Movement house found a welcome home at the seaside in the 1930s, there was still opportunity for new interpretations of traditional seaside colour, as at San Remo Towers, Boscombe (1936–8), a huge block of flats embellished with candy-striped faience. Although beach huts are little more than wooden huts with railed front verandas, having originated with the bathing machines of the Regency seaside, they can be endlessly decorative and have always been popular, despite occasionally being seen as deeply unfashionable. By the late 1990s they were in short supply in some resorts, and new 'huts' were built at Lowestoft, Brighton

The traditional beach amusement of the donkey ride is still available at Blackpool – although not on Fridays, when a local by-law gives the animals a day off; in the background is the North Pier. Camels were once tried as donkey substitutes, but Blackpool's wet summers were not to their liking.

The Egyptian House in Penzance was built around 1835 as a museum and geological repository. It is a rare surviving example of the style popular after Napoleon's 1798 campaign in Egypt; its Coade-stone ornamentation includes Egyptian figures and winged serpents. The house is now owned by the Landmark Trust.

Above: *Traditional seaside images on a roadside tile panel made by Carter's of Poole in the 1950s and designed by Arthur Nickols. These tubelined panels, showing yachts in Poole Harbour, once graced local approach roads and the Poole promenade. They were also made for other towns including Swanage and Sandbanks, where one is still extant in its original position near the harbour mouth.*

Above: *A la Ronde, near Exmouth, is a unique sixteen-sided cottage orné built in 1798 for Jane Parminter, daughter of a Devon merchant, and her cousin Mary Parminter on their return from a ten-year European Grand Tour; the site has superb views over Exmouth and the Exe estuary. The house, which was bought by the National Trust in 1991, is crowned by the Shell Gallery, an elaborate featherwork and shell fantasy room created over many years by the Parminters.*

Above: *A modern take on traditional beach-hut design at Lowestoft, where these colourful huts were built near the Claremont Pier, above the town's South Beach, during the late 1990s.*

Chalets on Princess Mary Promenade, Bridlington, built in 1998 as part of the South Promenade Improvement Scheme. Each of the thirty-seven chalets has individual access to the promenade and thence the beach by a 3 foot (1 metre) wide wooden bridge across the water channel.

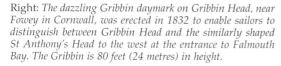

Above: *Workaday beach huts at Southwold; the lifeguards' (right) and fishermen's huts, with their strong colours and forms, display a stark contrast to the often humorous decoration shown by the typical holiday-maker's beach hut.*

Right: *The dazzling Gribbin daymark on Gribbin Head, near Fowey in Cornwall, was erected in 1832 to enable sailors to distinguish between Gribbin Head and the similarly shaped St Anthony's Head to the west at the entrance to Falmouth Bay. The Gribbin is 80 feet (24 metres) in height.*

(actually floating flats at Brighton Marina) and Bridlington, where high-quality chalets were included in the South Promenade Improvement Scheme.

The architecture of the working seaside, like the red-and-white-striped Gribbin seamark near Fowey and the black net lofts of Hastings, gave rise to 1950s definitions of the nautical style: strong, geometric form combined with solid, bright colour. Now, everyday structures including lifeboat stations and beach cafés contribute to good seaside design, while even such an unlikely candidate as Southern Water's waste-water treatment works at Ventnor (2000),

Peters Tower (right) was built at Lympstone, on the Exe estuary, in 1855 by William Peters as a memorial to his wife Mary Jane. The family, who lived nearby, gave the tower to the Landmark Trust in 1979. The tower, which carries a clock, was originally used as a refuge for fishermen; after restoration, and installation of a spectacular spiral staircase, the tower is now in use as a holiday home.

43

The Boston Deep Sea Fisheries office, opposite Lowestoft's harbour, was built by Consolidated Fisheries and known as Columbus Buildings. The white faience-façaded block has twin tiled panels of galleons – presumably representations of the Niña, the Pinta and the Santa Maria – at first-floor level; these brightly coloured, beautifully designed panels were probably manufactured by Doulton of Lambeth.

Left: The waste-water treatment works built by Southern Water at Ventnor, on the south coast of the Isle of Wight, was initially controversial because of its situation on the resort's beach just below the 1930s Winter Gardens. Perhaps its well-designed viewing gallery will eventually take the place of Ventnor's Royal Victoria Pier, demolished in 1993, as a promenading venue.

topped by an open circular viewing platform, has made a landmark out of a potential eyesore. From beach huts and Blackpool Tower to piers and promenades, a distinctively British style of seaside architecture has evolved, rooted in inventiveness and idiosyncrasy. It encompasses dynamic Victorian orientalism and cool 1930s modernism as well as today's hi-tech, artistically inspired structures and appears to be full of promise for the twenty-first-century seaside.

Southern Water's waste-water treatment works at Eastbourne could be a modern brown-brick interpretation of a Martello tower. The design of essential facilities such as these is much improved; the earlier Yorkshire Water pumping station on Bridlington's seafront was generally viewed as such an eyesore that it had to be re-clad when the promenade was improved in the 1990s.

Further reading

Adamson, Simon H. *Seaside Piers*. Batsford, 1977.
Bainbridge, Cyril. *Pavilions on the Sea: A History of the Seaside Pleasure Pier*. Robert Hale, 1986.
Boulton & Paul Ltd 1898 Catalogue. Algrove Publishing (Ottawa), 1998.
Braggs, Steven, and Harris, Diane. *Sun, Fun and Crowds: Seaside Holidays between the Wars*. Tempus Publishing, 2000.
Curtis, Bill. *Blackpool Tower*. Terence Dalton, 1988.
Fischer, Richard, and Walton, John. *British Piers*. Thames & Hudson, 1987.
Gray, Fred. *Walking on Water: The West Pier Story*. Brighton West Pier Trust, 1998.
Hix, John. *The Glasshouse*. Phaidon, 1996.
Ind, Rosemary. *Emberton*. Scolar Press, 1983.
Lindley, Kenneth. *Seaside Architecture*. Hugh Evelyn, 1973.
McLean, Bruce, Bauman Lyons, and Gooding, Mel. *Promenade: An Architectural Collaboration for Bridlington*. East Riding of Yorkshire Council, 2001.
Mickleburgh, Tim. *Glory Days: Piers*. Ian Allan Publishing, 1999.
Mickleburgh, Tim. *Guide to British Piers*. National Piers Society, 1998.
Pearson, Lynn. 'Tile and Tide: Seaside Ceramics', *TACtile: Newsletter of the Tiles and Architectural Ceramics Society*, 53–58 (2000–2).
Pearson, Lynn. *Amusement Machines*. Shire Publications, 1992.
Pearson, Lynn. *Lighthouses*. Shire Publications, 1995.
Pearson, Lynn. *The People's Palaces: Seaside Pleasure Buildings 1870–1914*. Barracuda, 1991.
Powers, Alan (editor). *Farewell My Lido*. Thirties Society, 1991.
Rutherford, Jessica. *The Royal Pavilion*. Brighton Borough Council, 1995.
Samuelson, Dale. *The American Amusement Park*. MBI Publishing (St Paul, Minnesota), 2001.
Saunders, David. *Britain's Maritime Memorials and Mementoes*. Patrick Stephens, 1996.
Turner, Keith. *Pier Railways and Tramways of the British Isles*. Oakwood Press, 1999.
Walton, John K. *The British Seaside: Holidays and Resorts in the Twentieth Century*. Manchester University Press, 2000.
Walton, John K. *The English Seaside Resort: A Social History 1750–1914*. Leicester University Press, 1983.
Walton, John K. *Fish and Chips and the British Working Class, 1870–1940*. Leicester University Press, 1992.
Ward, Colin, and Hardy, Dennis. *Goodnight Campers! The History of the British Holiday Camp*. Mansell, 1986.
Woodhams, John. *Funicular Railways*. Shire Publications, 1989.

The National Piers Society is a registered charity that aims to promote and sustain interest in the preservation and continued enjoyment of seaside piers. The Society can be contacted at 26 Weatheroak Close, Webheath, Redditch B97 5TF or 4 Tyrrell Road, Benfleet, Essex SS7 5DH. (NPS website: www.piers.co.uk).

Saltburn was faced with the possible loss of its pier in 1975, when the local council, owners of the pier since 1938, requested permission to demolish the structure. A public enquiry decided that only the far end of the pier should be removed, and it reopened (after restoration) in 1978; further restoration took place in 2000–1.

Places to visit

There are fifty-five British pleasure piers in existence. These piers are listed below, along with notable entertainment buildings and other interesting structures at resorts with and without piers; generally, lighthouses have not been included. Please note that mention in this list does not imply any form of public access, although many of these buildings are open to the public or visible from publicly accessible areas.

Cheshire
West Kirby: Column daymark (Column Road).

Cornwall
Fowey: Gribbin daymark (Gribbin Head). Penzance: Jubilee Pool, Egyptian House. St Ives: Tate St Ives.

Cumbria
Silloth: village.

Devon
Exmouth: A la Ronde. Ilfracombe: Landmark Theatre. Lympstone: Peters Tower. Lynton: Cliff Railway. Paignton: Pier, Festival Hall, Torbay Cinema. Plymouth: Tinside Pool. Teignmouth: Pier. Torquay: Princess Pier, Pavilion.

Dorset
Boscombe: Pier. Bournemouth: Pier. Branksome: Solarium. Swanage: Pier. Weymouth: Commercial Pier and Pier Bandstand.

Durham
Whitburn, South Shields: Souter Lighthouse.

Essex
Burnham-on-Crouch: Royal Corinthian Yacht Club. Clacton: Pier. Harwich: Electric Palace Cinema. Southend-on-Sea: Pier, Kursaal. Walton on the Naze: Pier.

Hampshire
Hythe: Pier. Southampton: Royal Pier. Southsea: Clarence and South Parade Piers.

Isle of Wight
Ryde: Pier, Pavilion. Sandown: Culver Pier. Totland Bay: Pier. Ventnor: Winter Gardens, waste-water treatment works. Yarmouth: Pier.

Kent
Deal: Pier. Dover: White Cliffs Centre. Gravesend: Pier, riverside café. Herne Bay: Pier. Margate: Dreamland.

Lancashire
Blackpool: Central, North and South Piers, Blackpool Tower, Winter Gardens, Casino (Pleasure Beach). Fleetwood: Pier, Radar Training Station. Morecambe: Victoria Pavilion, Midland Hotel. St Anne's-on-the-Sea: Pier. Southport: Pier.

Lincolnshire
Cleethorpes: Pier. Ingoldmells: original chalet (now office), Butlin's Holiday Camp. Skegness: Pier.

Penarth Pier was built in 1894–5, and a wooden pavilion added to its seaward end in 1907; the shore-end pavilion (which still retains its turnstiles) was built in 1927–8. The first pavilion and several shelters were destroyed in a fire on August Bank Holiday 1931. The pier was completely restored during 1994–8.

Norfolk
Cromer: Pier. Gorleston-on-Sea: Pavilion Theatre. Great Yarmouth: Britannia and Wellington Piers, Hippodrome, Winter Gardens, Gem (Windmill), Empire, Pleasure Beach.

Northumberland
Tynemouth: Railway (now Metro) Station, Life Brigade Watch House, seawater baths. Whitley Bay: Spanish City.

Somerset
Burnham-on-Sea: Pier. Clevedon: Pier. Weston-super-Mare: Grand and Birnbeck Piers, Knightstone Pavilion and Opera House.

Suffolk
Felixstowe: Pier. Lowestoft: Claremont and South Piers, East Point Pavilion, Royal Norfolk and Suffolk Yacht Club, Ness Beacon. Southwold: Pier. Thorpeness: village.

Sussex
Bexhill: De La Warr Pavilion. Bognor Regis: Pier. Brighton: Palace (Brighton) and West Piers, Royal Pavilion. Eastbourne: Pier, waste-water treatment works. Hastings: Pier, Cliff Railway, net lofts, White Rock Pavilion. Saltdean: Lido. Worthing: Pier, Dome Cinema.

Yorkshire – East
Bridlington: South Promenade, Floral Hall.

Yorkshire – North
Ravenscar: village. Saltburn: Pier, Cliff Lift. Scarborough: Grand Hotel, Cliff Lift, Northstead Manor Gardens, South Bay Pool, Spa Buildings.

Isle of Man
Douglas: Gaiety Theatre, Villa Marina. Ramsey: Queen's Pier.

Scotland
Dunoon: Queen's Hall. Rothesay: Winter Gardens, Pavilion, Pierhead Toilets. Wemyss Bay: Railway Station.

Wales
Aberystwyth: Royal Pier. Bangor: Bangor Garth Pier. Beaumaris: Pier. Colwyn Bay: Victoria Pier. Llandudno: Pier. Mumbles: Pier. Penarth: Pier. Portmeirion: village.

Museums
Beside the Seaside Museum, 34–35 Queen Street, Bridlington, East Yorkshire YO15 2SP. Telephone: 01262 608890. Website: www.bridlington.net/besidetheseaside

North Somerset Museum, Burlington Street, Weston-super-Mare, Somerset BS23 1LH. Telephone: 01934 621028. Website: www.n-somerset.gov.uk (Seaside Holiday Gallery)

The Column was erected at West Kirby, Cheshire, in 1841 by the Trustees of the Liverpool Docks. It took the place of an old windmill, destroyed by a gale in 1839, which had previously been used by mariners as a navigational aid. The stone-built daymark, largely a Doric column in form, overlooks the Dee estuary; there is an expansive view from its site on Caldy Hill.

Index